Walrus Facts and Myths

A Science Summary for All Ages

Susan J. Crockford

ISBN: 978-0-991-7966-6-3

One walrus. Ten walrus or ten walruses (both are correct).

A male walrus is a bull, a female is a cow and its baby is a calf.

A walrus cow feeds her calf milk just like a dairy cow does.

Acknowledgements

Special thanks to friends and colleagues who provided critical feedback and suggestions.

TABLE OF CONTENTS: FACTS & MYTHS

Acknowledgements v

1 Walrus are as big as a pickup truck 1

2 Walrus use their huge tusks to dig for food 3

3 Even newborn walrus calves have tusks 5

4 Walrus whiskers just keep growing 7

5 Walrus are found in all parts of the Arctic 9

6 Walrus are sometimes found outside the Arctic 11

7 Walrus live on sea ice in all seasons of the year 13

8 Walrus need to be fat to stay warm 15

9 Walrus like to sleep alone 17

10 Walrus are good climbers 19

11 Walrus have poor eyesight 21

12 Summer is the most important season for walrus 23

13 Arctic sea ice is melting 25

14 Less summer sea ice means more food for walrus 27

15 There are not very many walrus left in the world 29

16 Walrus are eaten by polar bears and killer whales 31

17 Walrus can get a sunburn lying on the beach 33

18 Climate change will harm walrus in the future 35

About the author (and answer to question p. 1) 37

List of photo credits 39

1. Fact or myth? Walrus are as big as a pickup truck.

Myth!

A walrus is as long as a small car but can be much heavier. See page 39 to find out what walrus-like animal *is* as big as a pickup truck. Can you guess?

A walrus is still very large – adult males can weigh almost 4,000 lbs (2,000 kg) – and measure 12 feet long (3.6 m). Female walrus are slightly shorter and usually weigh about half as much as males.

Walrus can rotate their back flippers forward to walk and climb on land. In the water, the big back flippers make walrus excellent swimmers.

Both male and female walrus have long canine teeth in the top of their jaws called tusks. Walrus, like elephant seals, have ears on the *inside* of their head for hearing but no ears *outside* that stick out or flap around.

The scientific name for the walrus is *Odobenus rosmarus*, which means 'tooth-walking sea horse' in Latin. Walrus can live for about 40 years.

2. Fact or Myth? Walrus use their tusks to dig for food.

Myth!

Tusks are used for fighting, as protection from polar bears, and for helping the walrus climb onto ice. The tusks grow as the walrus grows. The biggest, oldest males have the longest, thickest tusks.

The walrus in the picture below is using its tusks to steady itself while it breathes through a hole in the sea ice. Tusks are very strong but they *can* break, so some walrus end up with short stubby tusks.

In their jaws, walrus have simple, peg-like molar teeth that hold food in place until it can be swallowed. Walrus don't do much chewing.

3. Fact or Myth? Even newborn walrus calves have tusks.

Fact!

Tusks start to grow as soon as the walrus is born – you just can't see them! Young walrus have tusks you can see when they about two years old, like the calf in the photo above. What's different about its mother's tusks? Do you see that the tips have broken off?

Walrus tusks are slightly curved and can be up to two feet long (60 cm). This is pretty impressive but the tusks of an elephant can be much longer.

4. Fact or Myth? Walrus whiskers just keep growing.

Fact!

Whiskers appear before birth and grow quickly after the calf is born (see the picture above). The whiskers continue to grow slowly during its life but are worn shorter by feeding activities, including nursing.

The whiskers are short on the top of the snout and longer near the mouth. Each whisker near the mouth moves easily and is very sensitive, almost like a finger. Whiskers help the walrus find food.

The longer whiskers help the walrus sort potential food it finds on the ocean floor into 'good to eat' and 'not good to eat' – the whiskers help move the 'good' stuff towards the mouth.

5. Fact or Myth? Walrus are found in all parts of the Arctic.

Myth!

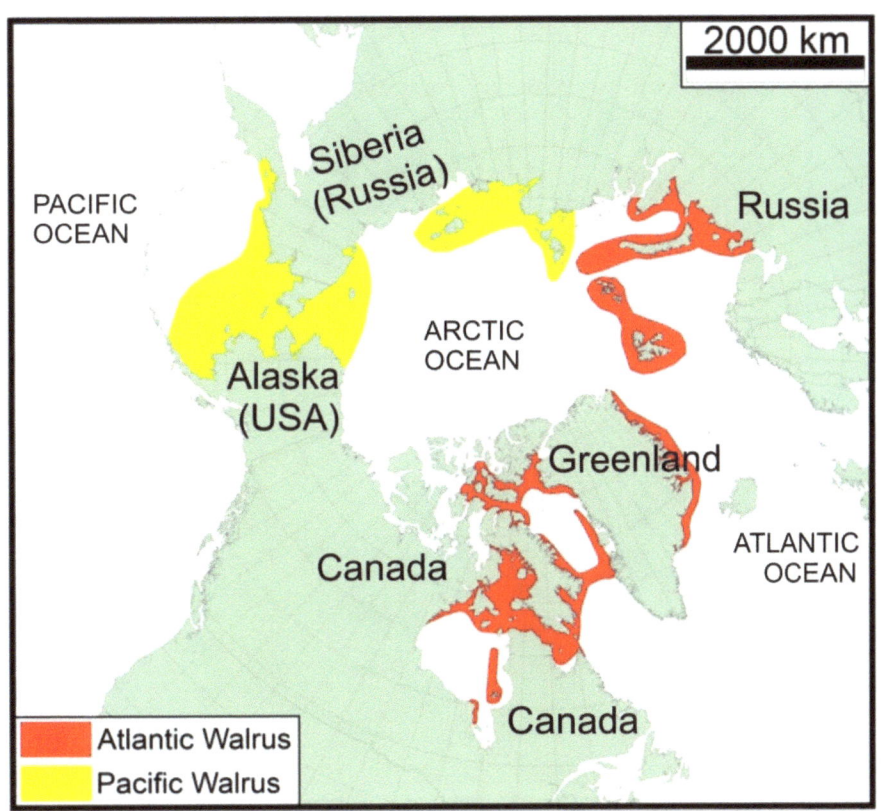

Walrus are divided into Atlantic (red on the map) and Pacific (yellow on the map) forms. Pacific walrus are slightly larger than Atlantic walrus and have longer tusks.

Since walrus need to dive to the bottom of the ocean to find their food, they are only found in Arctic areas were the sea is very shallow, near land.

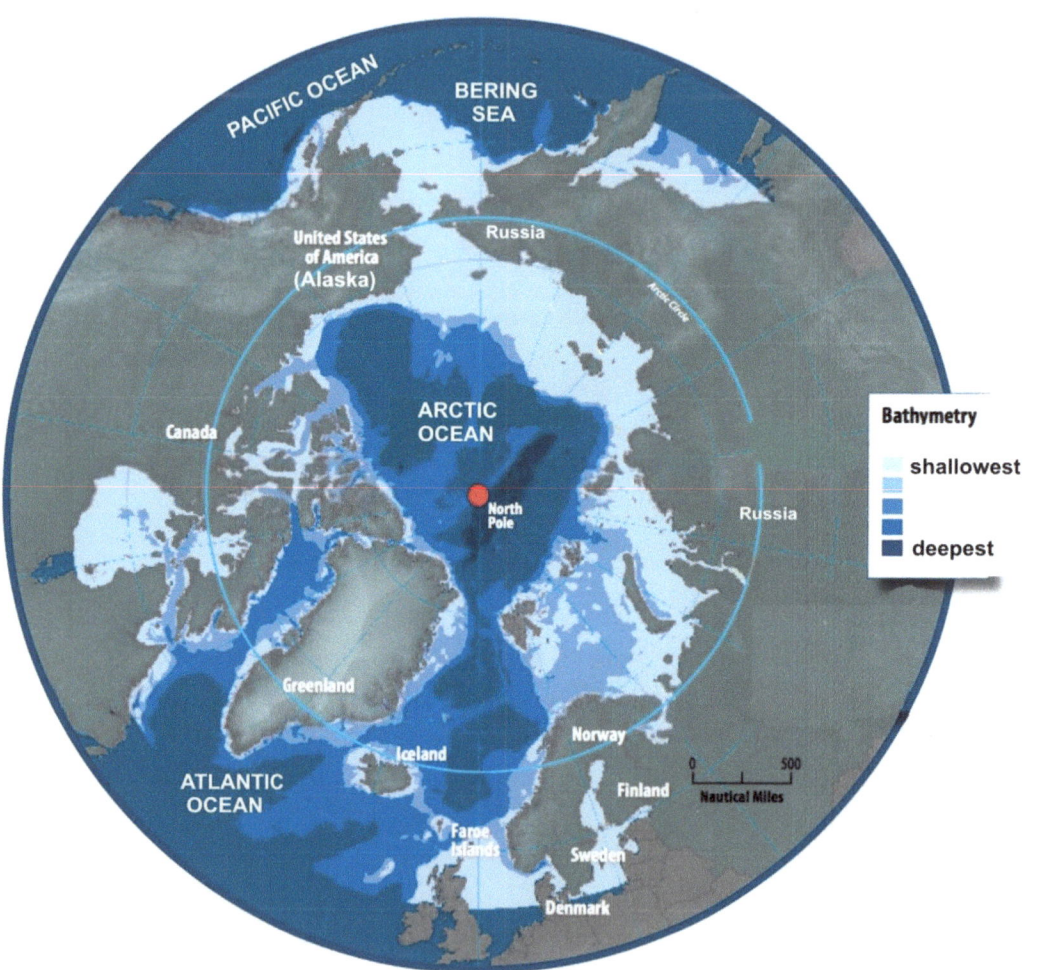

The light blue areas on this map of the Arctic are the most shallow and these are the places walrus prefer. Walrus eat mostly clams but also snails, sea cucumbers, worms, and even slow-moving fish. They suck clams from their shells and leave the shells behind.

6. Fact or Myth? Walrus are sometimes found outside the Arctic.

Fact!

Walrus that are almost full grown but too young to mate sometimes swim far outside the usual areas where older, mature adults live in the summer. Wandering walrus will sometimes leave the Arctic entirely.

In recent years, one Atlantic walrus was sighted as far south as Newfoundland, Canada and another in the Faroe Islands north of Scotland. In the Pacific, a walrus was spotted as far south as Japan.

Wandering walrus usually travel alone and stick to areas of shallow water so that they can feed easily on clams and snails on the bottom. The walrus above is resting along the coast of Newfoundland, Canada.

7. Fact or Myth? Walrus live on sea ice in all seasons of the year.

Myth!

Walrus use sea ice as a place to rest between dives for food and to give birth to calves, which are born in early spring (April or May). But in the summer, walrus often use ice-free beaches to rest between feeding. This has probably been true for millions of years but we know for sure it has been true for several hundred years.

In the spring, most male walrus leave mothers and calves and head to ice-free beaches, where they spend the spring and summer resting and feeding. Herds of mothers and calves also use beaches in summer and fall but often prefer sea ice.

Some Pacific walrus herds on shore have more than 100,000 animals. Above is a photo of about 1,500 Pacific walrus: most are on the beach but some are in the water returning from feeding. These beach resting areas are called *haulouts*.

8. Fact or Myth? Walrus need to be fat to stay warm.

Fact!

A thick layer of fat, called 'blubber', helps the walrus stay warm over the frigid Arctic winter when it must live on top of the ice. The sea water in the Arctic is very cold as well so being fat also keeps the walrus warm when it is swimming and diving for food.

However, like the polar bear, walrus also need stored fat to help them survive over the winter when it is very hard to find food.

Walrus need a mix of sea ice and open water so they can swim to the bottom to find clams: without some open water, they cannot dive.

9. Fact or Myth? Walrus like to sleep alone.

Myth!

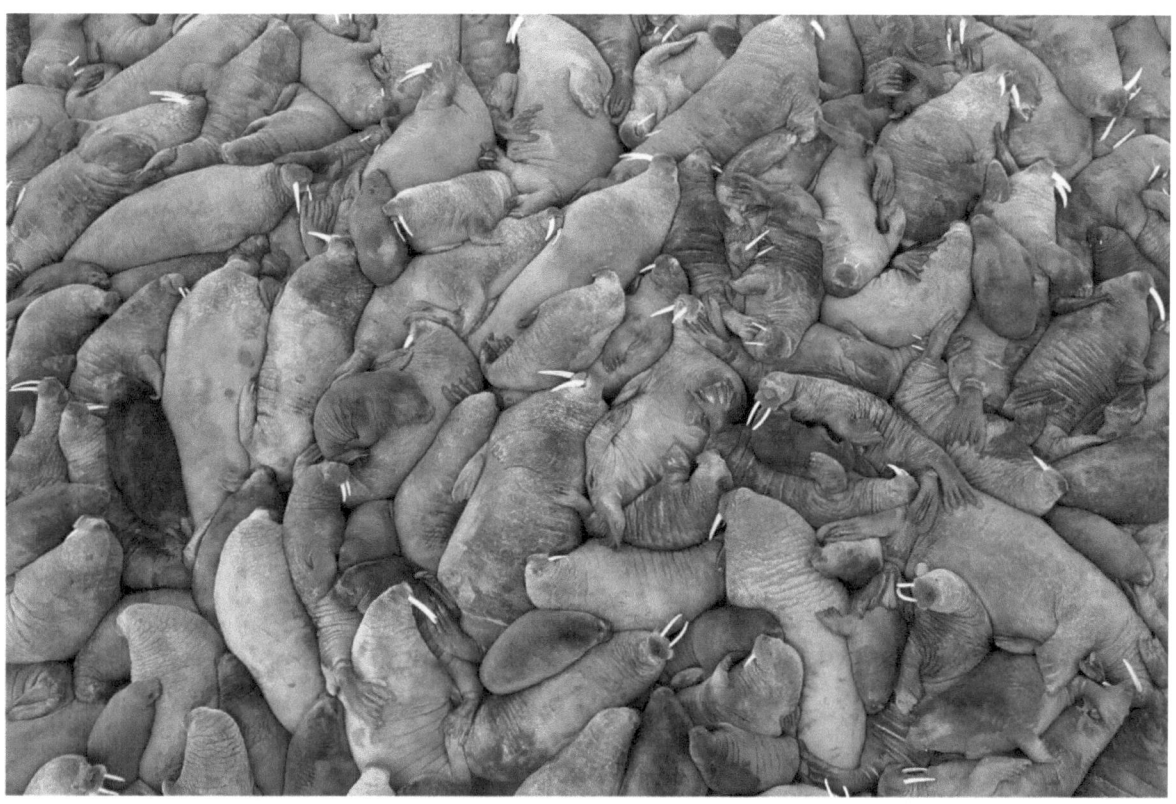

Walrus prefer to cuddle up to other walrus, even if there is room to spread out. On their backs with tusks in the air, on their sides, or on top of each other – they always seem to be relaxed and content.

This desire to crowd together means that young walrus have to work hard to stay out of the way when the huge adults move around and wiggle to get comfortable. If they don't watch out, the little ones could get squashed!

Walrus even pack in close when they are on the ice. They just seem to like being snuggled up together!

10. Fact or Myth? Walrus are good climbers.

Fact!

Walrus are excellent climbers. They need to be strong to pull themselves out of the water onto patches of sea ice and they use that strength to climb up onto rocky beaches.

Walrus cannot climb really tall cliffs like the one in the picture below but they can climb up less steep slopes very easily. But because they are so heavy, walrus have trouble getting *down* a hill without falling – they have go really slowly.

11. Fact or Myth? Walrus have poor eyesight.

Fact!

Walrus don't seem to be able to see very far but they have excellent hearing and a very good sense of smell. They would find it difficult to see someone standing still beside them at a beach haulout but are easily frightened by a sudden noise or something moving quickly nearby.

Walrus herds resting on a beach can be startled by noisy crows or flocks of seagulls flying over their heads. Sounds from airplanes and boats also scare them and so does the smell of a polar bear nearby.

Walrus on land head for the water when they are frightened by a sudden movement, sound, or smell. A walrus in the water that is scared will head for a patch of ice or a beach for safety.

12. Fact or Myth? Summer is the most important season for walrus.

Myth!

Spring is the most important season for walrus. Females give birth in early spring on the sea ice and use the ice as a platform for nursing their calves. As the sea ice starts to melt and break apart in late spring, all walrus eat a great deal after a long winter without much food.

Walrus also eat a lot during the summer but they don't need sea ice to do that. In summer, walrus can use beaches as a resting place between dives to the bottom to find food – or they can use a patch of ice as long as it is over shallow water.

In the fall, walrus cows and their calves migrate by swimming to areas where sea ice forms in the winter. Walrus bulls swim to meet the cows and calves, and they all spend the winter and early spring together on the sea ice.

13. Fact or Myth? Arctic sea ice is melting.

Fact!

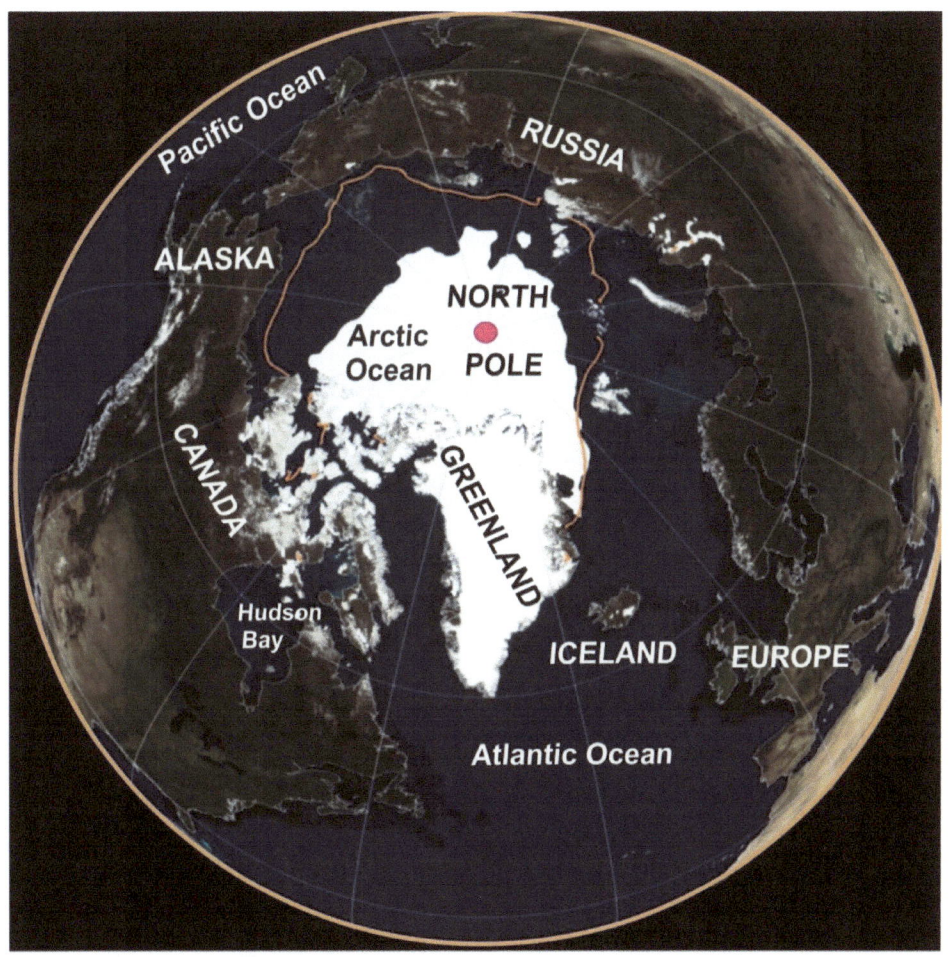

There is much less ice *in summer* now than there was 40 years ago. This map shows what the sea ice has been like in recent years at the end of September – ice covers only part of the Arctic Ocean instead of most of it, as it did 40 years ago (indicated by the thin orange line).

When people say that sea ice is melting, they really mean there is less *summer* ice than there used to be and that *summer* ice has more cracks and puddles.

Walrus have lived in the Arctic for millions of years and survived many periods of time in the past that had much more sea ice in winter than we have today – like the Last Ice Age – and times with much less ice in summer. These sea ice changes don't seem to bother the walrus.

14. **Fact or Myth?** Less summer sea ice means more food for walrus.

Fact!

Less ice in the summer means more sunlight reaches more areas of the Arctic and that makes life better for all sea creatures. That's because billions of tiny one-celled plants, called *phytoplankton*, live in the Arctic Ocean and they use sunlight and carbon dioxide to make food (*phyto* means 'plant' and *plankton* means 'made to wander or drift').

More sunlight means more phytoplankton. Phytoplankton is food for many other creatures in the sea, including the clams and snails that walrus eat. So more phytoplankton in summer means more food for walrus, seals, and other animals in the Arctic, including polar bears.

Walrus get very fat when there is lots of food to eat in spring and summer. A fat walrus is happy and healthy.

15. Fact or Myth? There are not very many walrus left in the world.

Myth!

Too much commercial hunting from the 1700s to the mid-1900s almost wiped out Atlantic walrus and drove Pacific walrus to very low numbers. Since then they have all been protected. Pacific walrus numbers today may be as high as they were one hundred years ago: the photo above is of a herd of about 35,000 walrus on a beach in Alaska.

Counting walrus is hard to do but scientists think that in 2016 there were probably more than 25,000 Atlantic walrus and more than 200,000 Pacific walrus.

What does 200,000 look like?

A one dollar bill is money but it's also a thin piece of paper that has a standard thickness – each one is the same size. If you had a one US dollar bill for every walrus and piled all 200,000 of them up, it would make a tower 860 inches tall – almost 72 feet (almost 22 meters).

That's the size of a 'B-Double' transport truck (a tractor-trailer with an extra trailer attached).

16. Fact or Myth? Walrus are eaten by polar bears and killer whales

Fact!

Polar bears eat mostly seals but they do eat walrus and giant whales that have died naturally. Walrus use their tusks to protect themselves from polar bears if they try to attack them face-to-face. So polar bears most often kill walrus by frightening a large herd resting on a beach or the top of a cliff and making them run for the water.

Frightened walrus stampede towards the safety of the water and in the process, small calves are sometimes trampled to death by bigger animals. Polar bears then eat the walrus that have died, like the bears in the picture below feeding on a dead whale.

Killer whales (also called 'orca') can catch walrus swimming in the water. With luck, the walrus will escape by climbing on a piece of ice or onto land.

17. Fact or Myth? Walrus can get a sunburn lying on the beach.

Myth!

The red you seen on some walrus on land comes from blood flowing just below the surface of the skin when the animals are warm. This is how a walrus controls its body temperature: the blood flows away from the skin into the body when it's in icy cold water or sitting on ice in the winter, and flows back towards the skin when it's warm out.

Walrus skin is very wrinkled and folded, even in very young animals. Older male walrus also have a number of thickened lumps all over their bodies called 'bosses'. Females don't have these.

Walrus skin is very thick, especially around the neck and shoulders. This helps protect the walrus from puncture wounds made by tusks during fights with other walrus.

18. Fact or Myth: Climate change will harm walrus in the future.

Myth!

In the Pacific half of the Arctic, walrus are thriving right now. In the Atlantic half, walrus numbers are rising steadily after they were almost hunted to extinction. Walrus have shown us that they know how to take care of themselves even if there is less summer ice than usual.

If ice disappeared completely in April or May sometime in the future, walrus might have trouble finding enough sea ice to give birth. But there is no sign of early spring ice disappearing, even in 100 years.

The Arctic is still very cold in the winter and spring and that means there is still a lot of sea ice when walrus really need it.

Answer to the question on page 1:

an adult male elephant seal

Elephant seal (male)

22 feet / 7,500 pounds

About the author

Susan J. Crockford has a Ph.D. in Zoology and has studied polar bear ecology and evolution for more than 20 years, and walrus ecology and conservation for seven years. She has authored many peer-reviewed papers about different animals (including polar bears) and has been writing a blog about polar bear science and walrus conservation since 2012. Her research interests include past and present aspects of Arctic biology and ecology.

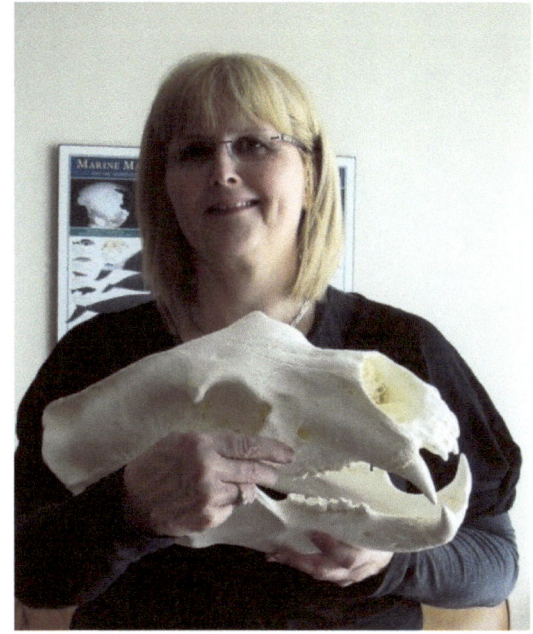

This book is the second in Dr. Crockford's *Facts & Myths* series. The first, *Polar Bear Facts & Myths,* is available in more than five languages, including French and German. She has also written a short, fully referenced science book about polar bears for adults and teens (*Polar Bears: Outstanding Survivors of Climate Change*) that's filled with useful color images in addition to more interesting facts. Her two most recent books for adults are *The Polar Bear Catastrophe That Never Happened* and *Fallen Icon: David Attenborough and the Walrus Deception.*

She is also the author of two science-based novels about polar bears that adults and older teens will enjoy: *EATEN* (a polar bear attack thriller set in Newfoundland, Canada) and *UPHEAVAL* (about a sea ice tsunami, set in Cape Breton Island, Canada – with a polar bear twist).

Website: www.susancrockford.com

Blog: www.polarbearscience.com

List of Photo credits

Front cover, Joel Garlich-Miller, US Fish & Wildlife Service, no date.

Back cover, Sarah Sonsthagen, US Geological Survey, July 2012.

ii Frontspiece, Shutterstock, purchased license.

iii Ryan Kingsbery, US Geological Survey, 2013.

iv Shutterstock, purchased license.

Pg 1, Shutterstock, purchased license and Go Graph purchased license (car).

Pg 2, Shutterstock, purchased license.

Pg 3, Shutterstock, purchased license.

Pg 4, Wikipedia Creative Commons license.

Pg 5, US Fish & Wildlife Service, 2011.

Pg 6, Wikipedia Creative Commons license and US Fish & Wildlife Service, 22 Nov. 2011.

Pg 7, Joel Garlich-Miller, US Fish & Wildlife Service, no date.

Pg 8, Sarah Sonsthagen, US Geological Survey, 2010.

Pg 9, COSEWIC 2017 Assessment and Status Report on the Atlantic walrus in Canada, labels added.

Pg 10, Wikipedia Creative Commons license, modified and labels added.

Pg 11, US Fish & Wildlife Service, no date.

Pg 12, Jim Walsh, Bay Bulls, Newfoundland, Aug. 2017.

Pg 13, Joel Garlich-Miller, US Fish & Wildlife Service, 2004.

Pg 14, Corey Accardo, US NOAA, 23 Sept. 2014, Pt. Lay, AK.

Pg 15, Shutterstock, purchased license.

Pg 16, Shutterstock, purchased license.

Pg 17, Reuters, still image from handout video, 3 Nov 2020, Kara Sea.

Pg 18, US Geological Survey, 2015.

Pg 19, Wikipedia Creative Commons license, Kolyuchin Island, Russia, 2013.

Pg 20, US Fish & Wildlife Service, Round Island, AK.

Pg 21, Ryan Kingsbery, US Geological Survey, 2013.

Pg 22, Joel Garlich-Miller, US Fish & Wildlife Service, no date.

Pg 23, Shutterstock, purchased license.

Pg 24, US Fish & Wildlife Service, no date.

Pg 25, US National Snow & Ice Data Service for Sept. 2012, labels added.

Pg 26, Shutterstock, purchased license..

Pg 27, Wikipedia Creative Commons license, Svalbard, Norway, 2013.

Pg 28, US Fish & Wildlife Service, June 2014.

Pg 29, Corey Accardo, US NOAA, 27 Sept. 2014, Pt. Lay, AK.

Pg 30, Wikipedia Creative Commons license (for both).

Pg 31, Shutterstock, purchased license.

Pg 32, Shutterstock, purchased license.

Pg 33, Shutterstock, purchased license.

Pg 34, Wikipedia Creative Commons license, Poolepynten, Svalbard, Norway, 2015.

Pg 35, Reuters, still image from handout video, 3 Nov 2020, Kara Sea.

Pg 36, Brad Benter, Bering Sea, US Fish & Wildlife Service, no date .

Pg. 37, NOAA Fisheries elephant seal/Ford trucks; Jesse McMillan, commissioned photo.